by Eileen Joyce

Printed in the United States of America

ISBN 978-0-15-362500-8
ISBN 0-15-362500-7

1 2 3 4 5 6 7 8 9 10 175 16 15 14 13 12 11 10 09 08 07

The Biosphere

Mention biodiversity, and most people will probably think of rain forests. Rain forests are, of course, one of the many environments on Earth's surface that support a large variety of life forms. But is the biosphere simply that thin layer of plants, animals, fungi, and other life forms that is spread out over Earth's surface? Perhaps we see life this way because we ourselves live on Earth's surface. And isn't that really where most life is?

In recent years, science has discovered life in surprising abundance in environments that are not technically on Earth's surface. Most of these environments should not, by conventional models, be able to support life. Some of these environments are below Earth's surface. They include the dark and sunless depths of the ocean, the boiling hot springs in Yellowstone, the ice in Antarctica, and the inside of rocks deep below Earth's surface where sunlight cannot penetrate and where air and food sources seemingly do not exist. The life forms that live in these environments are called extremophiles.

Extremophiles

Organisms that live in extreme conditions are called extremophiles. Of course, to an extremophile, its environment is not at all extreme, but the best possible environment for its adaptations. To an extremophile, our environment would probably seem too cold or too hot, or too filled with toxic gases, such as oxygen, to support life.

Most extremophiles are one-celled. Extremophiles are classified by the type of environment in which they live.

Some of the first extremophiles to be studied lived close to Earth's surface. In the early 1960s, a biologist named Thomas Brock was studying bacteria in Yellowstone Park's hot springs when he noticed signs of life in water that he had thought was too hot to support life. Near the outflow channel of Octopus Spring, he saw masses of pink filaments, or fine threads. The filaments were masses of bacteria thriving in water that was a scalding 88°C (190°F). When Brock and other scientists took a closer look, they found microorganisms in the boiling water of the hot springs. The microorganisms were not only surviving, but also thriving and rapidly reproducing. This was their home!

At first, the scientific community thought that these thermophiles (heat-lovers), as Brock had called them, were unusual. The scientists thought that Yellowstone's hot springs might have been an unusual sort of ecosystem that had forced a unique form of life to develop.

But then someone discovered life around the deep ocean vents.

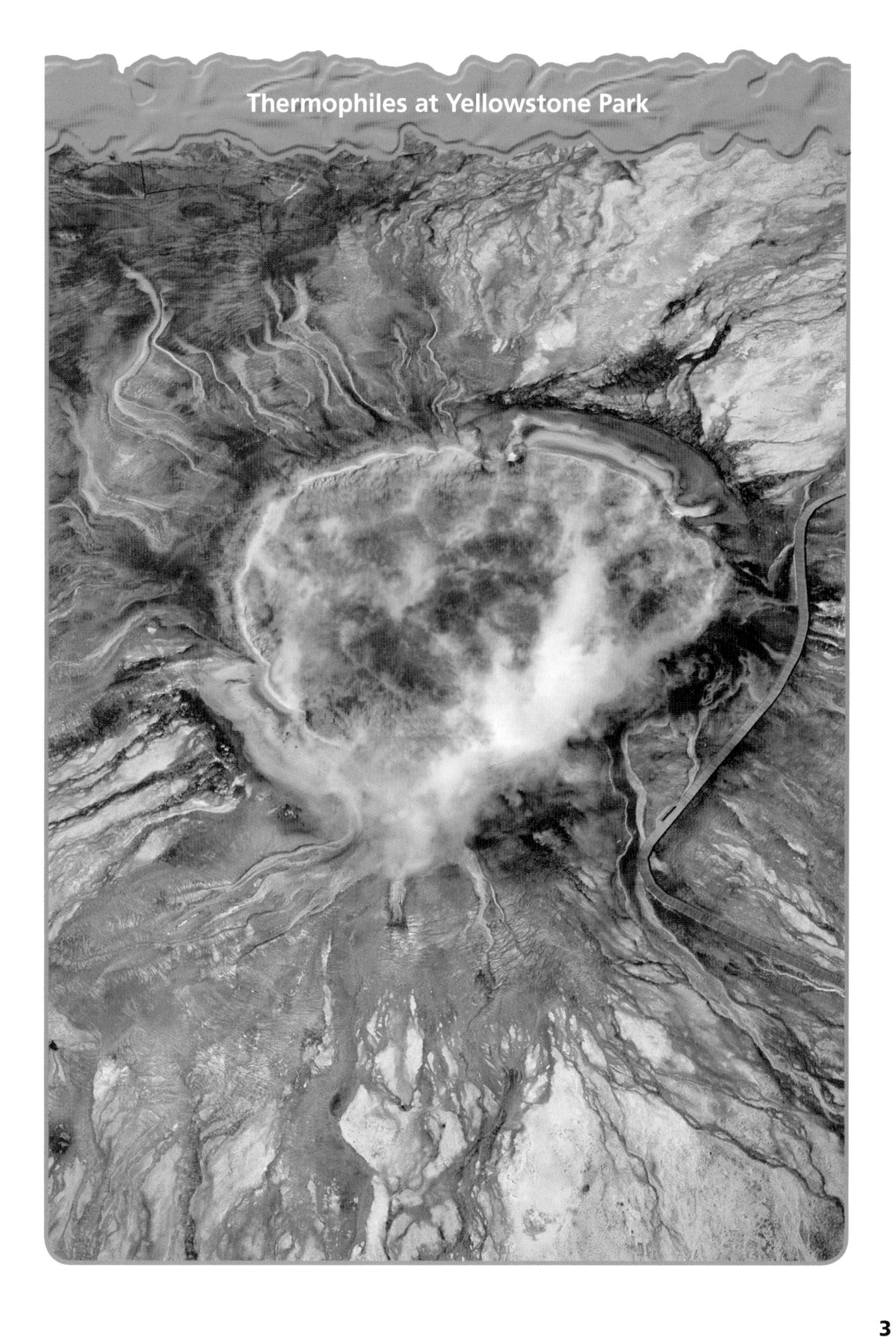
Thermophiles at Yellowstone Park

The Deep Biosphere

In 1977, scientists discovered hot springs in deep ocean water off the coast of Ecuador. The springs, called hydrothermal vents, were 2.5 kilometers (1.5 mi) down, on a spreading mid-oceanic ridge. Here the ocean's floor was slowly ripping apart and magma, at more than 1,000°C (1,832°F), was steadily erupting to form new oceanic crust. Scientists were not surprised to find the hydrothermal vents. They had been predicting that such vents would be found in areas of seafloor spreading. The surprise was the discovery of an entire community of organisms living around the vents.

Like the discovery at Yellowstone, the deep-ocean hydrothermal vent ecosystem found in 1977 was not unique. Many more hydrothermal vents have been found along mid-oceanic ridges in the years since then. The temperature of water around these vents can be as high as 380°C (716°F). For reference, water boils at 100°C (212°F). Nevertheless, this superheated water is home to an amazing number of organisms.

To find organisms living in such high temperatures was surprising enough, but the real puzzle for researchers was determining what the organisms fed on. Food chains on Earth's surface and even in the shallower parts of the ocean generally begin with organisms that can make their own food in a process called photosynthesis. On Earth's surface, we know most of these organisms as plants. Photosynthesis, however, depends on the presence of sunlight, and the organisms around the vents are far too deep for sunlight to reach them.

Scientists have since learned that the microorganisms that form the base of the ocean vent food chain use hydrogen sulfide to make their food in a process called chemosynthesis. Hydrogen sulfide is a chemical in the water that comes from the vents.

Larger organisms living around the vents eat the chemosynthetic microorganisms. Some of these microorganisms live inside the bodies of larger animals. One such larger animal is the tubeworm. The tubeworm has no digestive tract. It gets its nutrients directly from the chemosynthetic microorganisms living inside it.

Since the discovery of the deep ocean vents, many studies have been done on the unusual life forms around them. After finding and studying hydrothermal vents, scientists found something very similar, but also very different.

Hydrothermal vent creatures

Cold Seeps

Just as hot vents occur on the ocean floor, so do cold vents. These vents aren't really cold, but because they are not fed by geothermal activity, they are the same temperature as the surrounding ocean. Cold vents are more often referred to as cold seeps.

Like hydrothermal vents, cold seeps support a wide variety of chemosynthetic microorganisms as well as larger animals such as crabs and tubeworms. However, cold seeps are much less dynamic in nature than hydrothermal vents. Powered by rapid and almost explosive expansion, superheated water shoots from hydrothermal vents. Cold seeps, on the other hand, emit their fluids slowly and at a steady rate. As a result, the environment around cold seeps is more stable and the organisms live longer there than they do around hydrothermal vents. In fact, some researchers have concluded that cold seep tubeworms may be some of the longest-living invertebrates on Earth. Scientists estimate that cold seep tubeworms live from 170 to 250 years. Cold seep tubeworms also grow much longer than their hydrothermal vent counterparts. In the cold seeps, the tubeworms reach a length of about 2 meters (6 ft). They also grow down into the sediment of the cold seep, forming roots. It is through these roots that they absorb the chemicals that feed the chemosynthetic microorganisms living in their guts. Masses of tubeworms grow in bunches that resemble bushes. Like bushes, they provide places for other animals to live, hide, and feed.

Cold seeps were first discovered in 1984 in Monterey Bay, off the coast of California. Since then they have been found in many other parts of the world, including off the coast of Alaska, in the Sea of Japan, and in the Mediterranean Sea.

Scientists still aren't absolutely certain where the cold seeps' fluid comes from. Some think that it may be rainwater that has soaked through Earth's crust and made its way through cracks and holes to the edge of the continent. Others think that as the ocean-bottom sediments pile up, pressure squeezes out water, oil, and methane gas, causing these fluids to seep slowly out of the sea floor. Seeps are generally found at the edge of continents. Therefore, the fluid flow may be caused by tectonic processes.

Scientists do know, however, that cold seep environments have been around for a long time. About 60 fossil seeps have been found at various locations around the world. These seeps are thought to be about 500 million years old. The fossils indicate that life around the cold seeps looks about the same today as it did a half billion years ago. Scientists think that being at the bottom of the ocean protected these life forms from the pressures that drove other life forms to extinction or physiological change. Life forms closer to the surface were subjected to the effects of sudden climate changes, asteroid strikes, and any number of events that could have changed their environments. Meanwhile, at the cold seeps, life for the tubeworms and their neighbors went on as usual, relatively undisturbed.

A cold seep community

Deeper and Colder

No one expected to find so much life at the bottom of the deepest parts of the ocean. Just when they thought they had seen the strangest of the strange ocean-bottom life forms, scientists discovered methane hydrates at the bottom of the sea.

When we think of carbon fuels, also called fossil fuels, we usually think of oil, coal, and natural gas. But there is another type of carbon fuel that forms in areas of low temperatures and high pressures. Those conditions are found in sea-floor sediments and beneath arctic permafrost. In those conditions, methane, or natural gas, becomes trapped in ice crystals. Although the water and methane molecules do not react with each other to form a compound, they are tightly bonded together. It is as if the natural gas and ice form an entirely new substance. The substance looks like ice, but it burns if you put a lit match to it.

The scientific community was at first excited by having found such an unusual substance, and many photographs were taken of researchers holding burning ice cubes in the palms of their hands. But then, as additional fields of methane hydrates were uncovered, scientists realized that methane hydrates might contain more than twice the carbon found in all of the world's reserves of coal, oil, and natural gas. The suggestion was astounding. Methane hydrates could provide an energy source for many generations to come.

Because of their potential as an energy source, methane hydrates came to be closely studied by the scientific community. Then, in July 1997, came this announcement from the National Oceanic and Atmospheric Administration:

"A team of university scientists using a mini research submarine on an NOAA-funded research cruise has discovered, photographed, and sampled what appears to be a new species of centipede-like worms living on and within mounds of methane ice on the floor of the Gulf of Mexico, about 150 miles south of New Orleans."

Methane hydrates were actually a place where something lived! Like the living communities around seeps and hydrothermal vents, this was quite a surprise. Further studies indicated that the worms were living on chemosynthetic bacteria in the ice. Like the microorganisms around vents and seeps that made their food from sulfur compounds, these bacteria made their food from the methane in the ice. Scientists also found tiny shrimp that appeared to be feeding on the ice. These were similar to the shrimp that had been found at the vent and seep communities.

Methane hydrate: an ice that burns

Worms in the Glaciers

As strange as methane hydrate ice worms seemed to be, similar creatures had been found in the ice of a glacier more than a hundred years before. In 1887, a geologist named George Frederick Wright was exploring in Alaska when he reported finding ice worms living in glaciers there. Although the report was astonishing, it was nearly forgotten for 100 years. Myths and legends grew up about strange creatures that lived in the glaciers of the Yukon and other areas of the Northwest, but little more was said about the worms that Wright discovered until fairly recently.

When they first discovered ice worms, scientists thought that they might have been blown onto the ice from surrounding land surfaces. Studies demonstrated, however, that glacial ice worms spend their entire life cycles in the ice. They live in small pockets of water within the ice, near the glacier's surface.

Distant cousins to the common earthworm, ice worms are tiny, from a few millimeters to a centimeter (0.1 to 0.4 in.) in length, and a millimeter (0.04 in.) or less in diameter. Their scientific name is *Mesenchytraeus solifugus,* meaning "sun-avoiding worm." Like other extremophiles, the ice worms do not just tolerate the cold. They are entirely adapted to it.

An ice worm's perfect world has a temperature of about 0°C (32°F). If the temperature drops below -7°C (20°F) or rises above 4°C (40°F), an ice worm dies. When exposed to room temperature, ice worms disintegrate in about 15 minutes. To an ice worm, our environment is fatally extreme!

What do these worms eat? Unlike the hydrate ice worms, glacier ice worms have no chemosynthetic bacteria to graze on. A glacier ice worm's food source is at the glacier's surface. Therefore, to avoid the heat of the sun while eating, large numbers of worms go to the surface to feed after sundown. Once there, they eat red algae that live on the surface ice as well as pollen grains and fern spores that have been blown there by the wind.

It turns out that life in ice is not at all unusual. This fact was demonstrated again when bacterial life forms were found deep in the ice of the icy continent, Antarctica.

Life Under the Antarctic Ice

Antarctica is the only continent on Earth that is entirely inhospitable to human life. People who go to Antarctica must take their environment with them, building enclosures to protect them from an extreme climate with temperatures that never rise above a summertime high of about -30°C (-22°F). Although some larger animals, such as penguins, do live in Antarctica for short seasonal periods, the interior of the continent is a frigid and lifeless expanse of snow-covered ice reflecting the blinding light of the sun.

In the 1950s, geological surveys suggested that there could be lakes of liquid water far beneath the ice of this dry and barren area. Eventually, radar mapping done from airplanes and satellites indicated that there were about 80 such lakes beneath the glacial ice. Of these, Lake Vostok was the largest, with an area of about 10,000 square kilometers (6,213.7 mi) and a depth of about 600 meters (1,968.5 ft) at its deepest point. These measurements also made Lake Vostok one of the largest lakes on Earth.

Lake Vostok is not a frozen lake, but remains liquid as a result of a few combined effects. First of all, the lake is insulated from the cold by about 4 kilometers (2.48 mi) of solid ice. This ice sheet began to form over the lake about 15 million years ago, cutting it off from Earth's atmosphere.

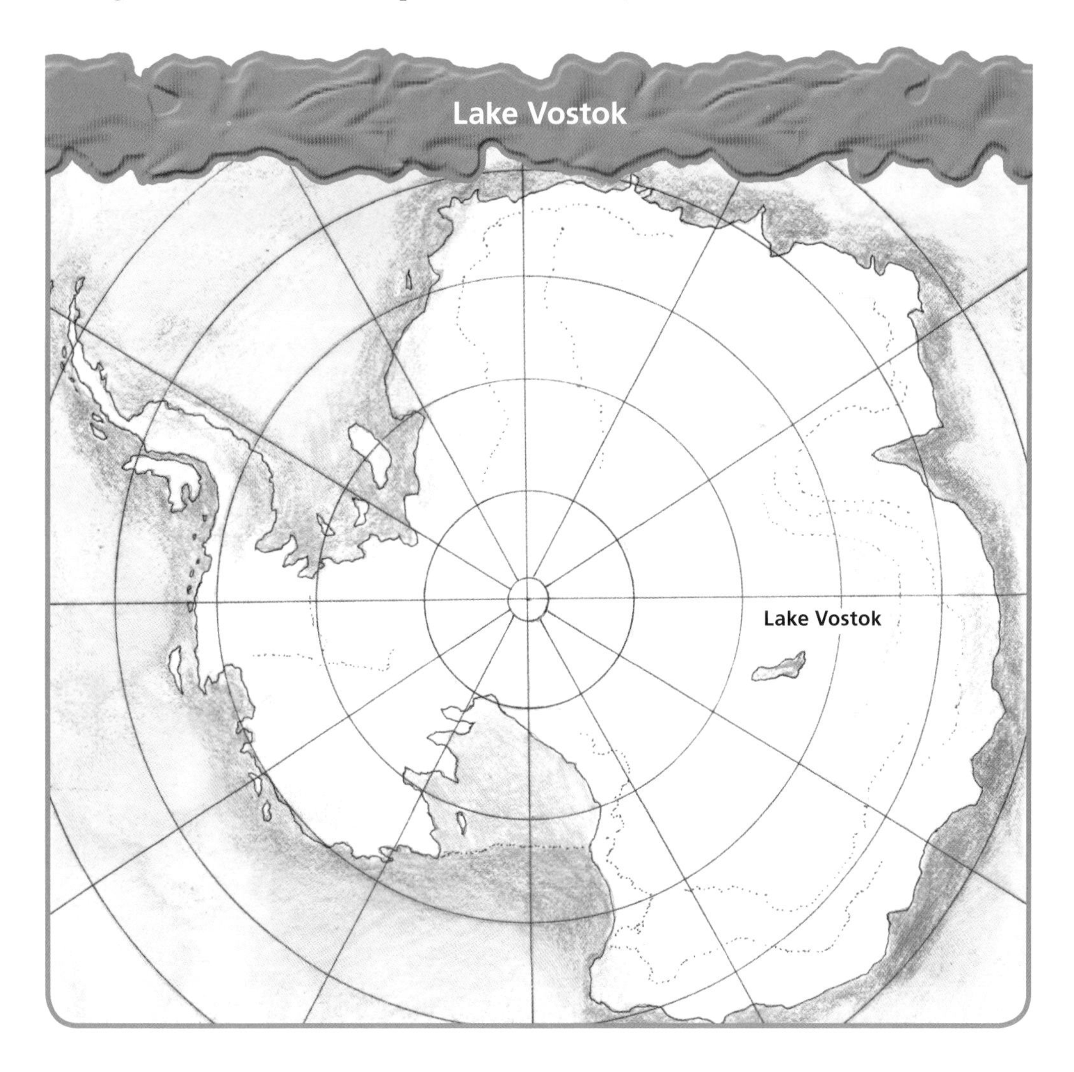

Beneath this insulating ice layer, the water is constantly warmed by the decay of naturally radioactive rock deep beneath the lake and by geothermal action. Like liquid in a pressure cooker with a tight-fitting lid, the water of Lake Vostok is believed to be under tremendous pressure as a result of the overlying ice and the underlying heat source. This pressure lowers the freezing temperature of the water, making the existence of liquid water even more likely.

In the late 1990s, scientists began to drill into the ice covering Lake Vostok. Although it would have been scientifically interesting to drill into the lake itself, and perhaps into the sediments underlying it, researchers were cautious. First of all, no one wanted to contaminate the water. Water in the lake had not been subjected to the

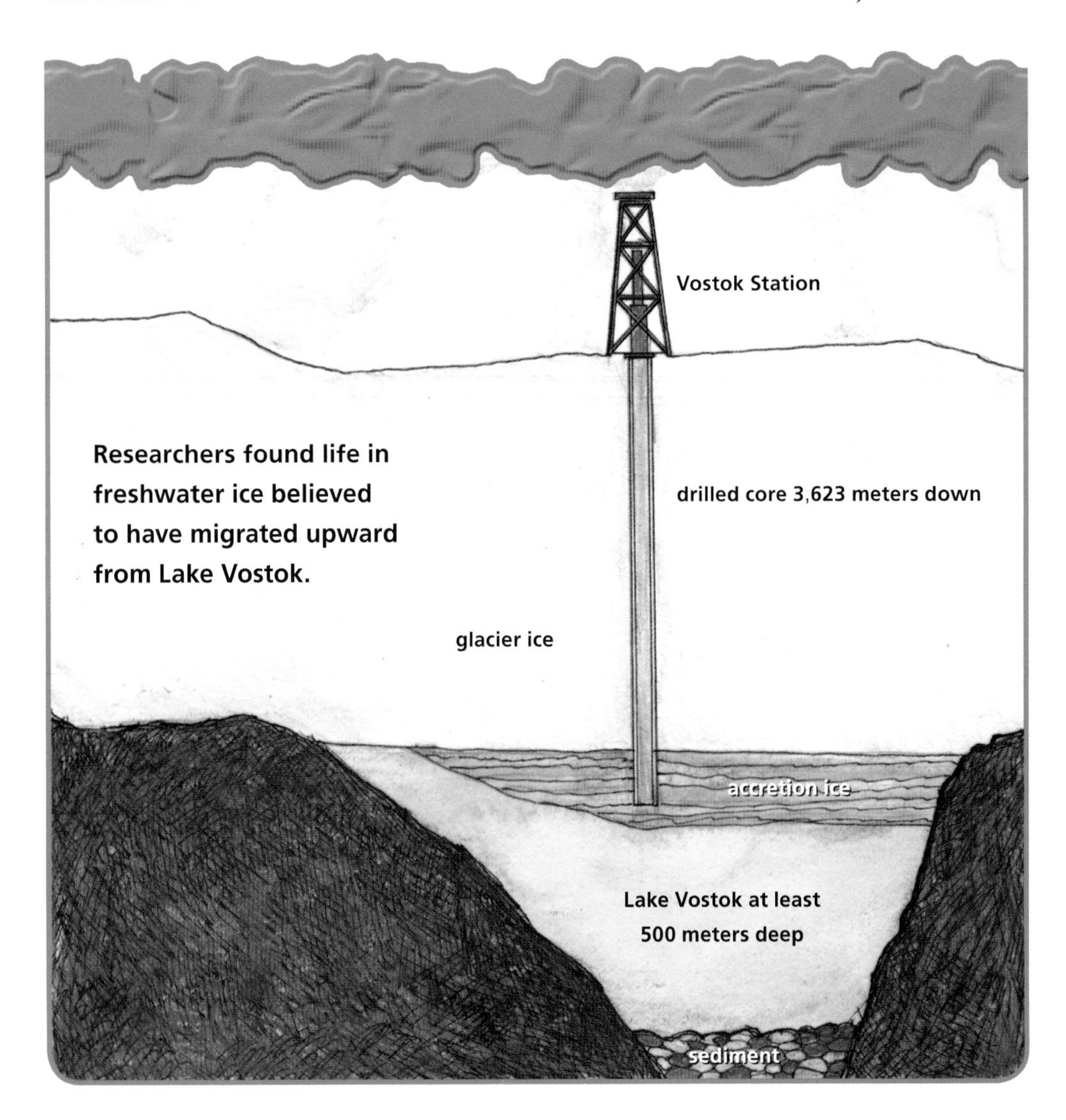

Researchers found life in freshwater ice believed to have migrated upward from Lake Vostok.

atmosphere for millions of years. An uncontaminated sample of the water might contain dissolved atmospheric gases from millions of years ago, giving scientists an opportunity to study the ever-changing makeup of the air. Also, if a sample of Vostok's water showed any trace of life, scientists wanted to be certain that the traces were from the lake itself rather than introduced by the drilling process.

A core sample was drilled down to about 3,600 meters (2.2 mi). The core revealed some surprising evidence about the dynamics of the ice. The ice sheet over the lake is in constant motion. As the ice moves over the lake, its underside picks up the surface water of the lake. This water refreezes as "accretion ice" in a separate layer of ice over the water.

One puzzling aspect of this constant removal and refreezing of surface water is how the moving ice sheet has not managed to empty Lake Vostok completely. Scientists think that new water is constantly flowing in from somewhere, but they do not agree on the source. It may be meltwater from the bottom of the ice sheet. The movement of the ice sheet also causes the water to circulate in the lake from bottom to top and back again.

Even more intriguing than the ice flow dynamics, however, was what scientists found in the ice. Microorganisms, including bacteria, yeasts, and fungi, were found throughout the ice core sample. Microorganisms in much of the overlying ice may have come from windblown dust when those portions of the glacier were still surface ice. But some of the microorganisms had been found in the accretion ice, that is, water that had been pulled from the lake and refrozen. Whether the accretion ice microorganisms had come from the lake's waters or from the overlying ice remains a question. But the presence of the microorganisms raises the possibility that Lake Vostok may indeed support life.

The likelihood of finding life in Lake Vostok seemed to improve in July 2004 when life was discovered in another lake beneath another glacier in another part of the world: Iceland. Grimsvötn Lake is about 100 meters (328 ft) deep and is covered by a 300-meter-thick (984.3 ft) ice sheet. Heat from geothermal processes beneath the lake keeps the water liquid. The environment is nonetheless cold and dark and is in many ways similar to that of Lake Vostok. If bacteria are alive in Grimsvötn Lake, then perhaps similar lakes are home to similar bacteria.

Until the water of Lake Vostok is probed, no one will know for certain whether the water contains living organisms. Scientists are interested in probing the sediments beneath Lake Vostok. This task may give them more of the history of how the lake formed, as well as any fossil evidence of life forms that have lived and died there. Scientists are also interested in the floor of Lake Vostok because of recent discoveries of life beneath the ocean.

Going Deeper into the Biosphere

In 1991, while diving in a submersible vessel, researchers witnessed the aftermath of an underwater volcanic eruption. The water seemed to be filled with white specks, almost as if the researchers were in a blizzard. The specks were sulfur and microbes. The ocean floor was covered with them. The microbes had not come from the ocean, but had been brought up from below the ocean floor by the eruption. Something was living even farther down, beneath the hydrothermal vent community.

The discovery raised a question: how far down into Earth's crust can life be found?

A blizzard of microbes

In 1999, researchers from Oregon State University found evidence of rock-eating microorganisms about a mile beneath the ocean floor, but they were unable to bring up live specimens. The researchers found tracks in the rocks that looked like passageways. "The microbes would make these little tubes…" said one of the researchers on the project. "They are either eating the rock or excreting some kind of acid that is doing it. One theory is that they are seeking [nutrients] in the rock."

In 2003, scientists from Oregon State announced that they had discovered living organisms 305 meters (1,000 ft) below the sea floor. The microorganisms were living in cracks in volcanic rocks.

It turns out that similar discoveries had been made beneath dry land many years before.

Ignoring and Reexamining the Evidence

It seemed impossible at the time. In the 1920s, a geologist from the University of Chicago found bacteria living in oil that had been taken from rocks 600 meters (1,969 ft) underground. Speculation that the bacteria lived deep beneath Earth's surface was rejected. The conclusion was that the bacteria had come from contamination of the drilling equipment.

Throughout the mid-1900s, similar evidence was also ignored. But then, in the 1980s, the United States Department of Energy began to study the ground deep beneath one of its nuclear facilities. They were concerned that if microbes did exist deep underground as rumors suggested, those microbes might affect the spread of chemical and radioactive contaminants. What they found was a diverse community of microbes living 500 meters (1,640 ft) underground.

The organisms were thermophiles, similar to those found in the hot springs at Yellowstone and around the hydrothermal vents in the deep ocean. The microbes found underground did not arrive there on the equipment that was used to dig down into the rock. It was natural for the microbes to be there. It was their home.

Today scientists estimate that life may extend down as far as 4 kilometers (2.5 mi) into Earth's continental crust and 7 kilometers (4.4 mi) into the oceanic crust. These figures suggest some fascinating ideas. If life can be found in the deep biosphere, without all of the necessities we take for granted on the surface, then it's possible that life may have begun there. In Earth's early days, after all, the surface was not the welcoming environment that we see today. It was being pounded by meteorites. Also, Earth's atmosphere offered no protection from the harmful untraviolet radiation of the sun. The safest place to be in those times was inside rocks, deep underground.

Extraterrestrial Biospheres

If life can thrive in such extreme environments as you have just read about, there is some possibility that it can exist on other planets or moons in our solar system. That idea is being explored right now. Two such places in our solar system where scientists are looking to find extraterrestrial life are Europa and Mars.

Europa is a large moon that orbits the planet Jupiter. Europa's surface is made of ice. Wrinkles, creases, and fractures in the ice indicate that liquid water may exist beneath the ice, warmed by volcanic action far beneath the water. In many ways, an environment on Europa might resemble the environments found beneath polar glaciers here on Earth.

But how do scientists get below the ice? Researchers are developing a probe that can melt through the ice to the water beneath. Called a cryobot, this probe is being tested on Earth's glaciers right now with some degree of success. The cryobot may get its first extraterrestrial test not on Europa, but on Mars.

Mars has glaciers, too. They are at both poles, forming permanent caps that grow and shrink with the seasons. The same type of environments found in and under glaciers on Earth may be found in or under the ice caps of Mars, as well. Researchers plan to send the cryobot to Mars in 2007.

There is no perfect environment for life. Scientists have realized this only with the discovery of a deep biosphere here on Earth. Finding life in extreme environments here on Earth may increase the likelihood of finding life in other unexpected places, both on Earth and elsewhere.